Meet the Bugs

By Brian Johnson

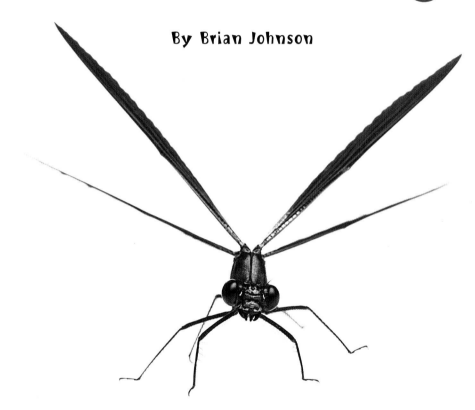

BRIGHT
connections media

A World Book Encyclopedia Company

Bright Connections Media
A World Book Encyclopedia Company
233 North Michigan Avenue
Chicago, Illinois 60601
U.S.A.

For information about other BCM publications,
visit our website at
http://www.brightconnectionsmedia.com
or call 1-800-967-5325.

Meet the Bugs
ISBN: 978-1-62267-004-8

Printed in China by Toppan Leefung Printing Ltd.,
Guangdong Province
1st printing July 2012

Acknowledgments
Front and back cover: © imagebroker/SuperStock; © Katrina
Brown, SuperFusion/SuperStock; © Exactostock/SuperStock;
© Shutterstock; © Thinkstock

Interior: © Shutterstock; © Thinkstock; © Alamy;
WORLD BOOK photo by Don Stebbing

Contents

Dragonfly

Have you ever seen a "horse stinger" or a "mosquito hawk"? You have if you've seen a dragonfly. These are all nicknames for the dragonfly.

Dragonflies have a lot of nicknames. They're often called darning needles, or devil's darning needles. People once believed that a dragonfly could use its thin, needlelike body to sew up a person's ears, eyes, and mouth!

Dragonflies have also been called horse stingers, because they often hover around horses and appear to be stinging them. But the real stingers are the flies the dragonflies are trying to catch! Dragonflies do not sting. You can hold a dragonfly in your hand or let it cling to your finger. It won't bite you—and it can't sting you.

But dragonflies do attack other insects. They're fierce, swift hunters. They've earned the name mosquito hawks because of the way they dart after mosquitoes, seizing them in midair. Dragonflies are also called bee hawks and bee butchers, because they kill and eat many honey bees.

Folk tales from the southern United States claim that dragonflies take care of snakes, so many people there call them snake doctors. This is because they think that dragonflies "look after" snakes! Some people even believe that a "snake doctor" can bring a dead snake back to life!

Ladybug

The plant leaf is covered with tiny green insects. These insects, called aphids (AY fihdz), suck juices out of plants. In time, they will kill this plant!

But, suddenly, a flying creature alights on the leaf and folds its wings with a click. It is a round-bodied, red-and-black ladybug.

The ladybug is quite a bit bigger than the aphids. Marching to the nearest aphid, the ladybug seizes it and gobbles it down! One after another, the ladybug eats the aphids until all are gone!

Ladybugs are very useful. They eat aphids and other insects that destroy many kinds of plants we use for food. In the late 1800's, a type of scale insect threatened to destroy the fruit crop in California. But the fruit was saved by turning thousands of ladybugs loose.

There are more than 5,000 different kinds of ladybugs in the world. Many are red or yellow with black spots. Some are black with red or yellow spots. Their bright colors are probably a "warning" to birds that ladybugs taste bad. And birds usually leave them alone.

The bright-colored back of a ladybug is really two hard wing covers. If you watch a ladybug getting ready to fly, you'll see its colored back split in two as it lifts its wing covers out of the way of its wings. Most kinds of beetles have wing covers. Wing covers are one of the things that make them beetles.

Tiger beetle

Many of the beetles you see scuttling about on a warm, sunny day are probably tiger beetles. They're called tigers because they hunt like tigers. They creep up on a smaller insect, then rush at it, seize it, and munch it down!

A mother tiger beetle digs tunnels and lays one egg in each. The larvae (babies) that hatch out look like long-legged caterpillars. Like their parents, they're meat-eaters. But they have a different way of hunting. After a larva hatches, it lies in wait just inside the entrance to its tunnel. When an insect passes close by, the larva rears out and grabs it. Then the larva pulls its captive into the tunnel and eats it!

There are many kinds of tiger beetles. Most of them are brightly colored and shiny. You can find some kinds on sandy beaches, some in woods, and some in gardens—where they help to keep out harmful pests.

Firefly

It is a summer evening, just at twilight. Long shadows fall across the grassy meadow and filter through the nearby forest. Suddenly, scores of tiny, flashing lights drift through the gathering darkness. The fireflies have come out.

Fireflies, also known as *lightning bugs,* are neither flies nor bugs. They are really beetles. You can often see them during the day, crawling or resting on leaves. At twilight, the males begin to fly and flash their lights. The females flash their lights in answer. The flashing lights are a signal that helps male and female fireflies find each other so they can mate.

A firefly's light comes from the underside of the firefly's body, near the back end. A firefly has special chemicals in its body. When these chemicals mix with oxygen, a gas that is in the air, the firefly is able to make a light. This light is a cold light—there is almost no heat. It is not like the lights we know how to make, which give off heat as well as light.

Firefly larvae hatch from eggs. These babies, which look somewhat like caterpillars, are fierce hunters! At night, they creep beneath loose earth and dead leaves in search of snails, worms, and the larvae of other insects. A firefly larva shoots poison into its prey with its mouthparts. The poison turns the creature's insides to liquid, which the larva then sucks out.

A few kinds of adult fireflies are also hunters. But many kinds don't eat at all. They live only to mate. Then they die.

Some kinds of female fireflies don't have wings. They look much like caterpillars. In some places, these wingless fireflies are called glowworms.

Pinching beetle

Two large brown beetles face each other on a log. They have their big, curved jaws, nearly as long as their entire bodies, wide open.

Suddenly, one beetle charges! The beetles bump together, struggling and pushing and clicking their jaws. Finally, one beetle tumbles backward and slips off the log. The fight is over. Nearby, a female beetle waits. The fight was over her! She would now be the winner's mate.

These big-jawed beetles are called *stag beetles*. This is because their jaws look like the antlers of a stag, or male deer. Only the males have these big jaws. Scientists aren't sure why because the jaws aren't good for much. They are so big and awkward that they often seem to get in the way. But they don't prevent the beetles from flying.

One kind of stag beetle is called a *pinching-bug* because the jaws look as if they ought to be able to give a good pinch. Actually, the big jaws of the males are quite weak. But the females, which have much smaller jaws, can really give you a pinch— one that will make your finger bleed!

Stag beetles look fierce, but they aren't. They feed mostly on sap that oozes out of trees. Stag beetle larvae look like fat white worms. The larvae eat the rotting wood of old logs and stumps, where the mother beetles lay their eggs.

Weevil

If you ever happen to see a beetle that looks as if it has a very long nose, it is probably a weevil. Because of their long "noses," weevils are also known as snout beetles.

The snout isn't a nose. It is actually a long mouth with jaws at the end. Having such a long mouth enables a snout beetle to chew deep into things. Mother snout beetles chew deep tunnels into nuts, fruits, seeds, and the stems and roots of plants. Then they lay eggs in the tunnels. When the larvae hatch, they have all the food they need.

The weevil family is the biggest of all the animal families in the world! There are at least 40,000 different kinds of these creatures. That's more than all the different kinds of birds, reptiles (snakes and lizards), amphibians (frogs and toads), and mammals put together!

Many kinds of weevils are pests. They spoil many kinds of fruits, nuts, seeds, grain, and vegetables. Some kill trees. One kind, called a boll weevil, does terrible damage to cotton plants.

But some weevils are a help to us. We use them to kill plants that are growing where they aren't wanted.

In 1919, the people of Enterprise, Alabama, erected a monument to the boll weevil. The beetle has so damaged area cotton crops that farmers were forced to plant other crops, for example, peanuts. In the long run, this saved the local economy. It is the world's only monument honoring an agricultural pest.

Grasshopper

The season is late summer, and a grasshopper is ready to lay eggs. At the end of her long body are several hard, sharp points. She pushes these into the ground, drilling into the dirt. From the end of her tail, she drops tiny eggs into the hole. A foamy, sticky stuff comes out with the eggs. It dries into a sponge-like covering, full of holes through which air can reach the eggs.

With a great leap, the mother grasshopper soars away. She will dig several holes and lay many more eggs. Many, many mother grasshoppers are also laying eggs, so there are tens of thousands buried in the field and fields like it.

In autumn, the green plants of summer begin to wither and dry. The adult grasshoppers are all dead. After a time, the ground freezes.

The cold of winter actually makes the tiny creatures inside the eggs begin to grow. This is nature's way of helping them. If the eggs simply needed warmth to hatch, they might hatch too soon, in late fall, rather than spring. Then the baby grasshoppers would be caught without food—grass, leaves, and grain available in the spring and summer.

Grasshoppers often cause terrible trouble! Sometimes enormous numbers hatch at about the same time. They begin to move across the land, eating as they go. Eventually, billions of grasshoppers take off in one gigantic swarm.

Where one of these swarms comes down, there is dreadful destruction! The billions of grasshoppers eat every green plant and leaf in their path. They can destroy hundreds of square miles of farm crops. These grasshoppers are called *locusts*. They are short-horned grasshoppers because they have short feelers, or "horns." No one knows why so many of them sometimes appear so suddenly.

Cricket

Crickets are the main music makers of the insect world. The steady *chirp chirp chirp* you often hear on a warm night is cricket music. Crickets make their music by rubbing the sharp edge on one wing against a row of bumps on the other wing.

There are many kinds of crickets. One kind, *house crickets,* likes to live in people's houses. Black, big-headed *common crickets* live in meadows and lawns, but they too often come into houses. *Tree crickets* stay in trees and bushes. There are crickets that live in caves, and crickets that live in ant nests—and are fed by the ants!

Most kinds of crickets are night animals. During the day they find a place to hide. At night they come out to eat. *Field crickets* eat grass, fruits, grain, and dead insects. House crickets will eat almost any kind of food. Tree crickets eat aphids, the tiny insects that live on leaves.

Cockroach

Would you like to see a live "prehistoric" animal—one that lived long before the dinosaurs? Well, look at a cockroach!

There were cockroaches swarming in hot, damp forests more than 300 million years ago! And they have hardly changed a bit since. They are what scientists call a "successful living creature." They have not had to evolve, or change, for millions of years. Many things have helped cockroaches survive. They are very fast runners. And their flat, slippery bodies allow them to quickly slide into cracks that most other creatures can't get into.

Cockroaches will eat almost anything—human food, garbage, dead plants and animals, cloth, leather, even wood. This has helped them survive. An animal that eats only one thing doesn't have as much chance to survive as does an animal that eats many things.

Mother cockroaches lay eggs that are protected inside a hard container that looks like a tiny purse. These containers can look like lumps of dirt, disguising them. Some cockroaches keep the containers with them until the eggs hatch. Other kinds give birth to live babies. All this helps keep cockroach eggs from being eaten by other creatures.

Most kinds of cockroaches live out of doors and never come near people. But, a few kinds do live in houses and buildings. They, too, hide during the day. At night they come out to eat. They get into people's food and spoil it. Because they can spread germs, cockroaches are troublesome, dangerous house pests. Most people dislike them, but no one has found a good way to get rid of them, except by spraying poisons.

Termite

In an old house, some of the wooden boards begin to crumble away. An old, dead tree suddenly splits and crashes to the ground. When such things happen, chances are that termites have been at work!

Termites are insects that eat wood. In North America, communities of termites live in the wooden parts of houses or in dead trees. The termites eat many long, branching tunnels in the wood around their nest. In time, the wood becomes almost hollow. What looks like a solid board may be as thin as paper!

Although termites eat wood, they can't digest it. Tiny, one-celled creatures that live in the termites' stomachs do this for them.

Termites are often called "white ants," but they aren't even closely related to ants. They're more like cockroaches. However, like ants, they live in communities.

At certain times, swarms of winged male and female termites fly from an old nest. When a male and female find one another, they dig a small, hidden nest.

After a time, the female lays her first eggs. The babies that hatch have six legs and can run about. In time, most become adult workers or soldiers. Some become winged adults.

The termite workers are small and have soft, pale bodies. Their small jaws are good only for digging or carrying things. But the soldiers are big, with large, hard heads and long, sharp jaws that are deadly weapons. Some kinds of termites have both workers and soldiers in the nest, but some kinds have only soldiers. Most workers and soldiers are blind. But they don't really need eyes. Most of them never leave the nest, which is completely dark.

Butterfly

A fuzzy, creepy, crawly caterpillar may turn into a butterfly or it might turn into a moth. Moths and butterflies are very much alike. They both start life as caterpillars. Often, only an expert can tell if a caterpillar will turn into a butterfly or a moth.

When a caterpillar starts to change into a butterfly or moth, it becomes what is called a pupa *(PYOO puh)*. Sometimes, you can tell if a caterpillar will be a moth or butterfly by the way it becomes a pupa.

Most butterfly caterpillars fasten their tail to a branch or stem and hang head down. A kind of shell forms around them. This shell is called a chrysalis *(KRIHS uh lihs)*. Inside the chrysalis, the crawly caterpillar becomes a butterfly.

Many kinds of caterpillars spin a covering of silk around themselves. This covering is called a cocoon *(kuh KOON)*. You'll often see cocoons on tree trunks and branches, on the sides of houses, and other places. Inside each cocoon is a growing caterpillar.

Butterflies live almost everywhere in the world. Tropical rain forests have the greatest variety of butterflies. Other butterflies live in woodlands, fields, and prairies. Some butterflies live on cold mountaintops. Others live in hot deserts. Many butterflies travel great distances to spend the winter in warm climates.

The largest butterfly, Queen Alexandra's birdwing of Papua New Guinea, has a wingspread of about 11 inches (28 centimeters). One of the smallest, the western pygmy blue of North America, has a wingspread of only about 3/8 inch (1 centimeter).

Scientists think that there may be about 20,000 to 30,000 species in total.

Moth

When caterpillars become butterflies or moths, it's still hard to tell which is which. Butterflies and moths both have the same kind of wings and body shape. Many moths are small, brown, and not very colorful—but so are many butterflies. Many butterflies are large, brightly colored beauties—but so are many moths.

Moths differ from butterflies in a number of important ways. For example, most moths fly at dusk or at night. The majority of butterflies fly during the day. Among most moths, the hind wing is attached to the front wing by a hook or set of hooks, called a frenulum *(FREHN yuh luhm)*. Butterflies lack a frenulum.

In addition, most butterflies have antennae that widen at the ends and resemble clubs. The antennae of most moths are not club-shaped. Most moths have antennae that either come to a point or that look like tiny feathers. No butterfly has such feathery feelers. Many male moths have larger antennae than do female moths.

Moths live throughout the world, except in the oceans. They inhabit steamy jungles near the equator. They have even been found on icecaps in the Arctic.

Moths vary greatly in size. The largest are the giant Hercules moth of Australia and the giant owl moth of South America. They have a wingspread of about 12 inches (30 centimeters). The smallest moths have wingspreads of about 1/8 inch (0.3 centimeter). These moths belong to a group called leafminers.

Ant

A female ant comes skimming out of the sky on long, rounded wings. Her black body shines in the sunlight. She lands gently in a meadow.

She started out in the morning, from a nest miles away—the nest where she was born. She will never see it again. Now she is going to start her own nest. She will be the queen and the mother. All the ants in the nest will be her children. While she was flying, she mated with several winged male ants from the old nest. Eggs are now growing in her body.

She begins to wiggle and rub herself against the ground. After a time, one of her wings comes off. She wiggles and rubs until the other wing breaks loose. She will never need wings again.

She begins to explore, trotting between the grass stems that rise like tall trees around her. Coming to a bare patch of sandy earth, she stops. Using her jaws and front feet, she digs. Soon, she has made a hole big enough to lie in. In the hole, she waits. Many months go by—nearly a year. In all that time, she never eats. She gets her food from her body fat and the two big muscles on her back, where her wings were attached. These muscles slowly shrink as her body turns them into food to keep her alive.

At last, the queen begins to lay eggs. She needs food badly—so she eats some of the eggs. Baby ants hatch from the others. The queen feeds them with a liquid that comes from her body.

After a time, the babies spin cocoons around themselves and change into adults. They are all females. But they have no wings, and they will never be able to lay eggs. They are workers—the first workers in the new nest.

The queen soon lays more eggs. Babies hatch. They are fed by the first workers. In time, the babies become adult workers, too. They are bigger and stronger than the first workers.

The tiny nest begins to grow. The workers dig tunnels and passageways in several directions. They make special rooms to keep the larvae in. The ants dig by grabbing bits of earth with their jaws and carrying the dirt up out of the nest. Then they go back for another bit. They do this again and again, until the tunnel or room is finished.

While some workers dig, others care for the larvae. They feed them with their saliva and move them about to keep them cool. Still other workers scurry out of the nest in search of food or scurry into the nest bringing food. An ant worker is always doing some job! No one "tells" the workers what to do. When one or two workers start doing things, other workers join in.

When an ant that is exploring outside the nest finds food, she hurries back to the nest. Soon, many ants stream out of the nest, straight for the food. The worker that found the food hasn't actually "told" the others where to find it. As she returns to the nest with a bit of food, she leaves a trail of scent. Other workers follow the scent trail made by the first ant.

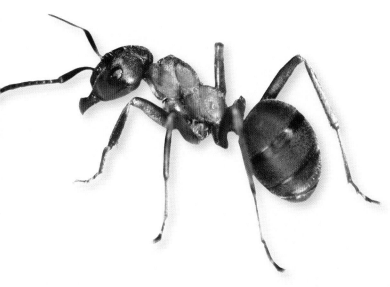

Smell is tremendously important to the ants. Every ant in the nest has the same smell, so every ant with that smell is a "friend." When a worker enters the nest with food, all the others know she belongs there. If another worker "asks" her for some food, she can tell by smell that the other is a friend. Then she'll put some of the food she carries in her mouth into the other's mouth. But any ant or other insect with a different smell is an enemy. It will be driven away if it tries to enter the nest.

All summer long the nest is a busy, active place. But in the wintertime it is still and silent. The ants lie motionless in the cold darkness. They are alive, but their bodies are too stiff to move. They are in a kind of "frozen" sleep. In the spring, the sun warms them, and they are able to move again. Once more, the ant colony bustles with activity.

Several years pass. Then, a very special day comes. Some of the queen's young have become winged females, as she once was. Others have become winged males. The time has come for these ants to leave the nest. It is like an ant holiday! As the males and females walk out of the nest, the excited workers swarm around them. One by one, the females and males soar into the air. There, they will mate. Then the females will land far away and start new nests. Thus, every ant nest is begun by an ant that has come from somewhere else.

Life in the old nest goes on for many more years. But one day the queen dies. So, no more eggs are laid. The ant colony is doomed. The workers die. The once bustling colony is empty and lifeless—like the ruin of an ancient city.

Honey bee

It is springtime, and an apple tree has burst into bloom. Its blossoms are puffs of white against bright green leaves. But a honey bee buzzing about nearby sees the blossoms as bright blue-green spots against a background of yellow leaves! The color and smell of the blossoms are a signal to the bee. This is how she knows the tree has something for her.

She alights on one of the blossoms and pushes down into it. Stretching out her tongue, which is like a long tube, she sucks up the bit of sweet nectar inside the flower. She flits from one blossom to another, sucking up more nectar. Soon, the little sack, or *honey stomach,* inside her body is full.

As she pushes into each flower, she is also doing something important for the tree. She gets a yellowish powder—pollen—all over her hairy body. Some of this pollen rubs off in each flower she visits. Pollen from one flower makes seeds start to grow in another flower. In this way, the bee helps the flowers make seeds from which new apple trees will grow. This process is called "pollination."

The bee also collects some of the pollen for food. She scrapes some of it off herself and mixes it with a tiny bit of nectar, to make sticky lumps. She pushes these golden lumps into "baskets" of stiff hairs on her back legs.

When her "baskets" and honey stomach are full, she heads straight home, making a beeline for her hive. The hive, which is in a hollow tree, is home to some 50,000 bees. Most of them are workers. Although all workers are females.There is but one queen bee. She does nothing but lay eggs.

The worker bees search for food, keep the hive clean, care for the young, and guard the entrance to the hive. To protect herself and defend the hive, each worker has a stinger on her tail. When a worker stings a creature, hooks on the end catch in the creature's body and the stinger is pulled out of the bee. A worker bee dies a few hours after she loses her stinger.

Reaching the hollow tree, the bee that had been at the apple tree goes past guard bees and through a crack in the trunk. Hanging inside are a number of honeycombs, long, thick strips of wax. Each comb is made up of thousands of tiny "rooms," or cells, all joined together. Thousands of workers built these combs out of bits of wax that came from their bodies.

The six-sided cells are storerooms. Many are filled with honey or with pollen—food for the bees. In some cells there are long, slim eggs. In others, there are little, wormlike baby bees. Still other cells are covered with wax lids. In these cells, baby bees lie wrapped in silk cocoons. The baby bees are changing into adults.

The bee that found the apple tree alights on part of a comb that has many empty cells. At once, she is surrounded by other workers. Several stretch out their tongues to her. She brings up nectar from her honey stomach. Each one sucks up some of it. But they don't eat it. One worker scurries off to feed the nectar to a baby bee. Others put their nectar into empty cells. In a few days, it will thicken into honey.

Meanwhile, the bee that brought the nectar begins an odd sort of "dance." She does her "dance" over and over again, while the other workers swarm excitedly around her. She is showing them where the apple tree is. Each time she makes a straight line, she is pointing the direction to the tree. And the speed at which she makes circles shows the distance to the tree.

Soon, many workers are winging toward the apple tree. Some fill themselves with nectar. Others collect only pollen. When the "baskets" on their legs are filled, these bees fly back to the hive. There, they scrape the pollen into empty cells in the honeycombs.

When a cell is filled, other workers take over. They let wax ooze out of their bodies. They chew it until it is soft and then use it to make wax covers for all of the pollen-filled cells.

Workers hatch from most of the queen's eggs. But a few baby queens hatch. And some males hatch. Soon, these babies start to spin cocoons around themselves. This seems to be a signal. Suddenly, the queen and many of the workers rush out of the hive! They fly away in a great swarm. They will start a new hive somewhere else. Never again will they return to the hollow tree.

Bumble bee

Bumble bees are usually bigger and sturdier than honey bees. They live in an underground nest, which is often in an old mouse hole. Like honey bees, they collect nectar and make honey. But they don't store the honey in wax combs. They save their old cocoons and store their honey in them.

A bumble bee queen starts a nest by herself in the spring. She makes a tiny "pot" out of wax from her body. Into this, she puts a ball of pollen and nectar. She lays eggs in the pollen ball. When the baby bees hatch, they're lying on their food.

The queen lays eggs during the summer. Most of the babies become workers and gather food. But not much food is needed. At summer's end, all the bees die.

During late summer, young queens and males hatch from some of the eggs. They fly off and mate. In winter, young bumble bee queens hide in the ground. In the spring, they come out and life starts again.

Unlike honey bees, bumble bees do not lose their stinger or die after using it. Bumble bees can sting again and again.

Wasp

Wasps are any of a large number of insects closely related to bees and ants. There are actually more than 17,000 species of wasps. Most have wings and fly.

There are two types of wasps: solitary and social. Common solitary wasps include mud daubers, cuckoo wasps, and digger wasps. Common social wasps include hornets and yellow jackets.

Female wasps have a stinger hidden near the end of the abdomen. Solitary wasps use their sting to paralyze prey. Social wasps use their sting to defend their nest.

Adult wasps feed mainly on the nectar of flowers. They prey on other insects and on spiders chiefly to provide food for developing offspring. Most species of wasps hunt a particular kind of prey. For example, one wasp species, the bee wolf, preys on honey bees. In most species of solitary wasps, the female captures and paralyzes prey by stinging it. She then puts it in a nest and lays an egg on the animal's body. When the larva hatches, it feeds on the prey.

Wasps are talented nest-builders, and each type has different nesting habits. Most solitary wasps dig nesting burrows in the ground. Other solitary wasps build nests completely of mud. Females do all the work.

Social wasps have a social system in which members of the community build and maintain the nest together. Most social wasps make their nests of paper. The female produces the paper by chewing up plant fibers or old wood. She spreads the paper in thin layers to make cells in which she lays her eggs. Some species of wasps, including a group called Polistes, build open nests with a single comb of cells. Other species, such as hornets and yellow jackets, construct nests of many cells enclosed by a paper covering with a single entrance. The nests may be suspended from trees, or they may be built underground in abandoned rodent burrows. A few kinds of social wasps build delicate mud nests.

Cicada

Every year, some part of the eastern half of the United States is invaded by red-eyed monsters!

The creatures appear in springtime or early summer. On a night in May or June, they push up out of the ground by the tens of thousands. By morning, the ground is dotted with holes.

These creatures are brown insects with big red eyes. Many people call them *17-year locusts*. But they aren't really locusts. Locusts are grasshoppers. These are *cicadas (suh KAY duhz)*. Cicadas are relatives of the leafhopper. Because they appear regularly after a certain period of years, scientists call them *periodical cicadas*. However, there are so many groups of young cicadas that one or more groups appear somewhere every year.

The instant a cicada leaves the ground, it heads straight for the nearest tree. Soon, all the nearby trees are filled with the little creatures. If there are no trees, the cicadas will climb onto bushes, weeds, or even blades of grass. When the cicadas come out of the ground, they're not quite grown-up. They go into the trees and stay there until they are adults.

Each cicada clings to its perch and humps its back. The skin splits down the back, and a more mature cicada struggles out. It is yellowish-white, with two black spots on its back behind its red eyes. Slowly, two ragged bumps on its back swell up to become soft, white wings.

Over time, the cicada's body turns gray, then black. The wings become clear, shiny, and yellow, with orange veins. The cicada is now an adult and ready for life. For a female, this means mating and laying eggs. For a male, it mostly means making a lot of noise— a steady buzzing song that attracts females.

Stink bug

Stink bugs are the skunks of the insect world. When a stink bug is in danger, it sprays a bad-smelling and bad-tasting liquid out of openings near its back legs. This keeps a bird or other animal from eating it.

There are many kinds of stink bugs. Most kinds have bodies shaped like the shields used by knights of old. In the United Kingdom, stink bugs are usually called *shieldbugs*. Many kinds of stink bugs are brightly colored, but some look like bits of bark. Some are the color of leaves.

A great many kinds of stink bugs are harmful pests that spoil vegetables and fruit. But some stink bugs are helpful. They eat harmful caterpillars and other insects.

Stink bugs are members of the group of insects scientists call *bugs*. We often call any insect or tiny creature a bug. But true bugs are a particular group of insects. All true bugs have a sharp beak for a mouth.

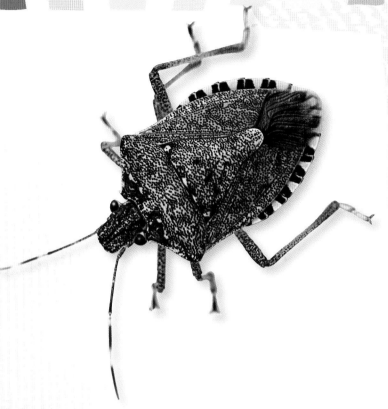

Some bugs are tiny, and some are large. Some have wings; some do not. Most bugs live on land, but some live in water. Some true bugs, such as stink bugs and water bugs, have "bug" as part of their name. But other "bugs," such as June bugs and potato bugs, are not true bugs but members of the beetle family.

Mosquito

A cluster of slim, oblong objects—some 200 eggs of a single female mosquito *(muh SKEE toh)*—floats on the surface of a pond. About two days after the eggs have been laid, the first baby pushes its way through its egg. It is a slim, wormlike creature. With quick jerks of its body, the larva swims away. Poking a tube at the rear of its body up out of the water, the larva takes in air, then wriggles to the bottom for its first meal of tiny plants and animals.

For about a week, it eats, grows, and sheds its skin. After the fourth shedding, it changes into a pupa. Hanging under the surface of the water, the pupa's skin splits after two days. A male mosquito climbs out and takes to the air. He eventually pokes into a flower and sucks up a meal of nectar.

The females leave the pond after the males. Eventually, the males join in a "dance." Beneath them, a swarm of females also join in a dance. From time to time, a female rises up and flies straight toward a male. Together, they leave and glide downward to mate.

After mating, the male lives for only a week. Living longer, the female will eventually feed by jabbing a tube through the skin of an animal or human to suck blood. She then lays eggs. She will continue to take more meals of blood and lay more eggs. After about a month, she will die, and the whole cycle starts over.

Mosquitoes spread germs and parasites that cause some of the world's deadliest diseases. The common house mosquito transmits such viruses as the Saint Louis encephalitis virus and the West Nile virus. Some mosquitoes transmit malaria, which kills millions of people each year in Africa and parts of Asia.

House fly

What may be the most dangerous animal in the world to people? A shark? A tiger? A snake? It's none of these. It's the common house fly! Why? Because flies carry germs.

Female house flies lay up to 250 eggs at a time, usually on garbage or rotting food. The babies start feeding on the garbage as soon as they hatch. Called *maggots,* they are tiny, white, legless, and wormlike.

After five or six days, a maggot's skin becomes a thick, brown shell. Inside, the maggot changes into an adult fly with two wings.

Because a fly tastes with its feet, it usually walks all over things "trying them out." In this way, it gets germs on its feet. When it walks on people's food, it leaves germs on the food.

Flies also spread germs when they eat. A fly can only suck liquids. But, it can turn some things into liquids. It does this by dissolving them with a liquid it "spits up" from its stomach.

A fly may suck up a liquid that has germs in it. This puts the germs into its stomach. If the fly then "spits up" on, for example, the sugar on the dinner table, the germs get onto the sugar. If you eat the sugar, the germs get into you.

But not all flies are house flies. There are many kinds of these two-winged insects. Many are bothersome, but not all are dangerous. Some are even helpful—they carry *pollen* (a fine, yellowish powder) from one flower to another, helping new plants grow.

Trap-door spider

Some kinds of spiders live in an underground burrow, or tunnel, with a door or lid that opens and closes. The door even has hinges on it! These spiders are known as *trap-door spiders*.

A young trap-door spider digs a tunnel in the earth. It lines the tunnel with silk—like a room covered with thick curtains. Then the spider makes a round door out of layers of dirt and silk. This is fastened to the entrance. The door is thick and fits into the entrance like a cork in a bottle.

The outside of the door is made to look like the ground all around the tunnel. When the door is closed, the entrance can hardly be seen.

A male trap-door spider leaves his nest only once in his life, to move in with a female and mate. A female seldom leaves her nest. The nest is an *ambush*—a place from which the spider can make a surprise attack on its prey.

When a trap-door spider is hungry, it lurks at the tunnel entrance and holds the door slightly open so that it can peep out. Then it lies in wait until an insect comes by. When an insect comes close enough, the spider reaches out and grabs it. The spider gives the insect a bite with its poisonous fangs, to paralyze it. Then it drags the insect into the nest and eats it.

But some of the spider's enemies, such as wasps, are able to chew through the door. Then the spider has to fight for its life. Some kinds of trap-door spiders don't depend on a thick door to save them. Their front door is made mostly of silk and is thin. But these spiders have a trick they play if an enemy breaks through the door. They have a small side tunnel, branching off the main tunnel. This side tunnel has a door that can be pulled shut to close it off from the main tunnel. The door blends in perfectly with the tunnel walls. An enemy is often tricked into thinking the tunnel is empty!

One kind of trap-door spider uses part of its body as an extra "door." The spider's back part is very large, hard, and flat across the end. If an enemy breaks down the spider's front door, the spider turns around and goes down its tunnel, headfirst. The tunnel is narrower near the bottom, and the spider's back part is soon wedged in tightly. It forms a thick, hard "door" that an enemy can't bite, sting, or chew through.

Dangerous spiders

Most spiders are harmless and even rather helpless. They will almost always scurry out of your way unless forced to defend themselves. Although many spiders have a poisonous bite, the bite of most spiders won't hurt you. However, the bite of some larger spiders can be as painful as a bee sting. And the bite of a very few kinds of spiders can cause pain, sickness, and even death.

The *black widow spider,* which lives in North America, can be dangerous. It often makes its web in the corners of garages, barns, or sheds. The northern black widow spider is black with red marks on its back.

The *redback spider,* which lives in Australia, is related to the black widow. This spider is also dangerous. It is black, with a broad red stripe on its back.

Another dangerous North American spider is the *brown recluse (REHK loos* or *rih KLOOS).* It sometimes hides in closets, drawers, or under furniture. It is brown, with a black mark shaped like a violin on its back.

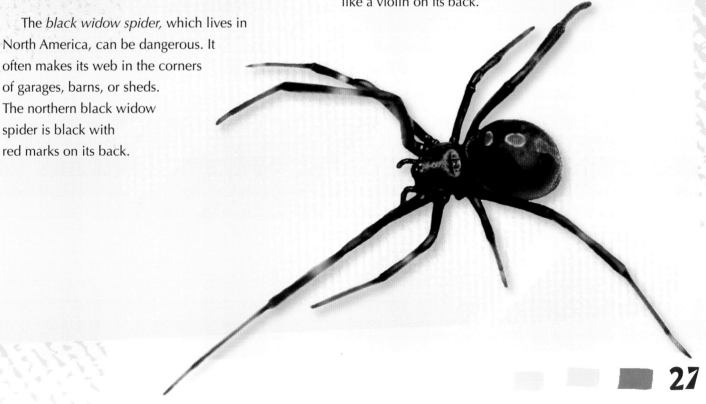

Spider giants

The biggest spiders are a family of furry, black or brown creatures that live in warm parts of the world. In North America, they are usually called tarantulas (*tuh RAN chuh luhz*). In other places, they are known as hairy spiders or bird-eating spiders.

In South America, some of these spiders are large enough to cover a dinner plate! They are hunters. They creep over the ground or through the branches of trees and leap on their prey. They usually hunt insects, but they also catch and eat birds, lizards, frogs, mice, and other creatures much larger than they are!

Most people are afraid of such large spiders. A few are dangerous, but most are not—their bite is no worse than a bee sting. Some kinds can be tamed and even make good pets!

Longlegs

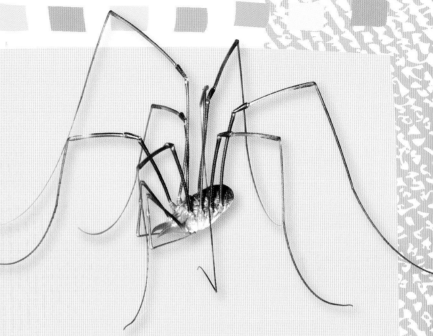

In late summer, or early autumn, you'll often see the creature called a *harvestman*. It got this name because so many of them are seen at harvest time. Usually, harvestmen move about slowly in shady spots. They seem to glide along on their eight enormously long legs, which are as thin as hairs. Because of these long legs, most North Americans call this harmless creature daddy longlegs.

Harvestmen are "cousins" of spiders. But they're different from spiders in many ways. A spider's body has two main parts. The head and body of a harvestman, however, form a single, egg-shaped torso. Unlike spiders, which have many eyes, a harvestman has only two eyes. It cannot make silk as a spider can, and it doesn't have poisonous fangs as a spider does. Instead, it has a pair of claws by its mouth.

Most kinds of harvestmen eat tiny insects that they tear to pieces with their claws. They'll also eat any large, dead insects they come across. Some kinds of harvestmen can break open snail shells with their strong claws. They then dig out the snail,

bit by bit. And some kinds of harvestmen drink the juice of fruit by squeezing the fruit with their claws.

Harvestmen defend themselves against some enemies by squirting a bad smell out of openings in their body. And if a bird grabs a harvestman by one of its long legs, the leg comes off. It wiggles as if it were alive and keeps the bird busy while the harvestman escapes.

Female harvestmen lay their eggs in the ground, under stones, or in cracks in tree bark. They lay their eggs in autumn. The babies hatch the following summer. They usually stay in hiding most of the spring, while they are growing up.

Centipedes and millipedes

A little, wormlike animal known as a slug lies in the darkness under a flat stone. If it could think, it would probably think it was quite safe. What could get at it under the stone? But another creature is also under the stone—a centipede *(SEHN tuh peed)*, and it is hungry. With its many feet and long, flat body, it moves quickly and easily under the rock. Its feelers jerk and twitch. One of them touches the slug.

In an instant, the sharp fangs on the centipede's jaws stab the slug's body. Poison squirts through the hollow fangs, and the slug becomes dinner.

There are many kinds of centipedes. All are fast and fierce, hunting in dark places. They prey mostly on insect larvae, spiders, cockroaches, worms, and slugs.

Most centipedes are small. But the giant desert centipede found throughout the southern United

Centipede

States and Mexico is very large. Its prey includes large insects, lizards, mice, and even birds.

A centipede's narrow body is divided into many sections. Each section has one pair of legs. The name centipede means "hundred feet." But almost no centipede has exactly 100 legs. Most have about 70 legs. Some centipedes have only 30, and others have 360! A centipede's head includes two antennae (jointed feelers) and a pair of jaws. The first pair of legs behind the head are modified into fangs. The fangs are called poison claws because a gland in the head fills them with poison. The bite of some centipedes can be dangerous to human beings.

Centipedes, like many other insects, hatch from eggs. Some kinds of baby centipedes look just like their parents. Others have only a few legs when they hatch. Each time they shed their skin, they gain more sections and more legs.

One kind of centipede likes to live and hunt in people's houses. Most people try to kill a centipede if they see one in the house. But they shouldn't. A house centipede is harmless. And it can be helpful. It will keep the house clean of roaches, flies, silverfish, and other creatures that are harmful.

Millipedes *(MIHL uh peedz)*, like centipedes, generally live under logs, rocks, and piles of leaves. But their way of life is very different from that of a centipede. Millipedes crawl about very slowly. And they aren't hunters. They eat mostly bits of dead, rotting plants. But some species also attack the roots of crops growing in damp soil. This causes problems for gardeners and farmers.

Millipedes defend themselves with poison. They give off a smelly liquid that has a stinging, burning, bitter taste. A few species are capable of producing fluids containing cyanide poison. Birds or other animals that try to eat a millipede get a real taste shock. A millipede also has a hard shell, which protects it when it curls up in a circle.

Like centipedes, millipedes hatch from eggs. Most kinds of millipedes make a nest for their eggs. The female builds a hollow dome out of dirt she mixes with liquid from her mouth. She lays from 20 to 300 eggs through a hole in the top of the dome. Then she seals up the opening and goes on her way.

The young millipedes hatch in several weeks. Like centipedes, the body of a millipede has many sections. Two pairs of legs attach to most sections. At first, they have only a few pairs of legs. As they get older, they grow more sections and more legs. They have round heads that bear a pair of short antennae.

The name *millipede* means "thousand feet." And some people do call millipedes "thousand-legged worms." But no millipede has this many legs. The greatest known number of legs is about 750.

There are about 7,500 known kinds of millipedes. They live in all parts of the world. But there are probably many kinds of millipedes that scientists have not yet discovered.

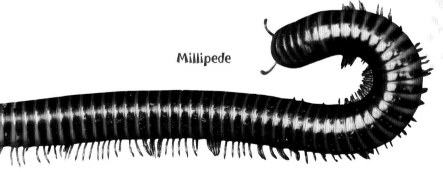

Millipede

Walkingstick

Many people have seen what they thought was part of a bush suddenly start to walk! What they thought was a twig sticking up from a branch was really an insect that looks like a twig. The strange appearance of this insect hides it from its enemies.

Most people call these skinny, twiglike insects *walkingsticks*. Many kinds of walkingsticks can't jump or fly, because they have no wings. They live mostly in trees or bushes.

Most kinds of walkingsticks in North America are only about 2 to 3 inches (5 to 8 centimeters) long. But in Australia there is a walkingstick about 10 inches (25 centimeters) long. And in Indonesia there is a type of walkingstick more than a foot (30 centimeters) long!